BEYOND 2000
MICROMACHINES
AND
NANOTECHNOLOGY
The Amazing New World of the Ultrasmall

BEYOND 2000
MICROMACHINES AND NANOTECHNOLOGY

The Amazing New World of the Ultrasmall

by David Darling

Dillon Press
Parsippany, New Jersey

Acknowledgment

The author is grateful to Professor Stephen D. Senturia, the senior researcher in micromachines at the Massachusetts Institute of Technology, for his expert comments and advice during the preparation of this book.

Photo Credits

Cover: © Peter Menzel

Ferranti Ltd./University of Manchester: 15, 18. © Seth Joel/Science Photo Library/Photo Researchers, Inc.: 14. Professor Mehran Mehregany, Case Western Reserve University: 8. © Peter Menzel: Table of Contents & Chapter 1-6 Opener (Insets), 30, 55. William McLellan: 52 Musee d'Art et d'histoire, Neuchatel (suisse): 12. © David Parker/Seagate Microelectronics Ltd./Science Photo Library/Photo Researchers, Inc.: 23. © Sinclair Stammers/Science Photo Library/Photo Researchers, Inc.: 40. © ChrisTaylor/CSIRO/ Science Photo Library/Photo Researchers, Inc.: 45. © Sheila Terry/Rutherford Appleton Laboratory/Science Photo Library/Photo Researchers, Inc.: 20. Illustrations by Marie T. Dauenheimer: 6, 26, 32, 35, 38, 43, 47, 50.

Library of Congress Cataloging-in-Publication Data

Darling, David J.
    Micromachines and nanotechnology: the amazing new world of the ultrasmall/by David Darling.—
    1st ed.
        p. cm.—(Beyond 2000)
    Includes bibliographical references and index.
    ISBN 0-87518-615-7   ISBN 0-382-24953-4 pbk
  1. Nanotechnology—Juvenile literature. [1. Nanotechnology.] I. Title. II. Series.
    T174.7.D37   1995
    620.4—dc20   94-26595

Summary: Describes current and possible future developments in the manufacture of machines of extremely small size, including the technology used and the important ways in which these tiny, moving devices may be applied.

Copyright © 1995 by David Darling

All rights reserved. No part of this book may be reproduced or transmitted in any form or by any means, electronic or mechanical, including photocopying, recording, or by any information storage and retrieval system, without permission in writing from the Publisher.

Published by Dillon Press, an imprint of Silver Burdett Press.
A Simon & Schuster Company
299 Jefferson Road, Parsippany, NJ 07054

First edition

Printed in Mexico

10 9 8 7 6 5 4 3 2 1

CONTENTS

Introduction		7
Chapter 1	The Incredible Shrinking Machine	10
Chapter 2	Computers, Crystals, and Chips	17
Chapter 3	Gearing Down	28
Chapter 4	How Small Can You Get?	37
Chapter 5	The Atomic Erector Set	44
Chapter 6	Amazing Nanotechnology	51
Glossary		57
For Further Reading		61
Index		62

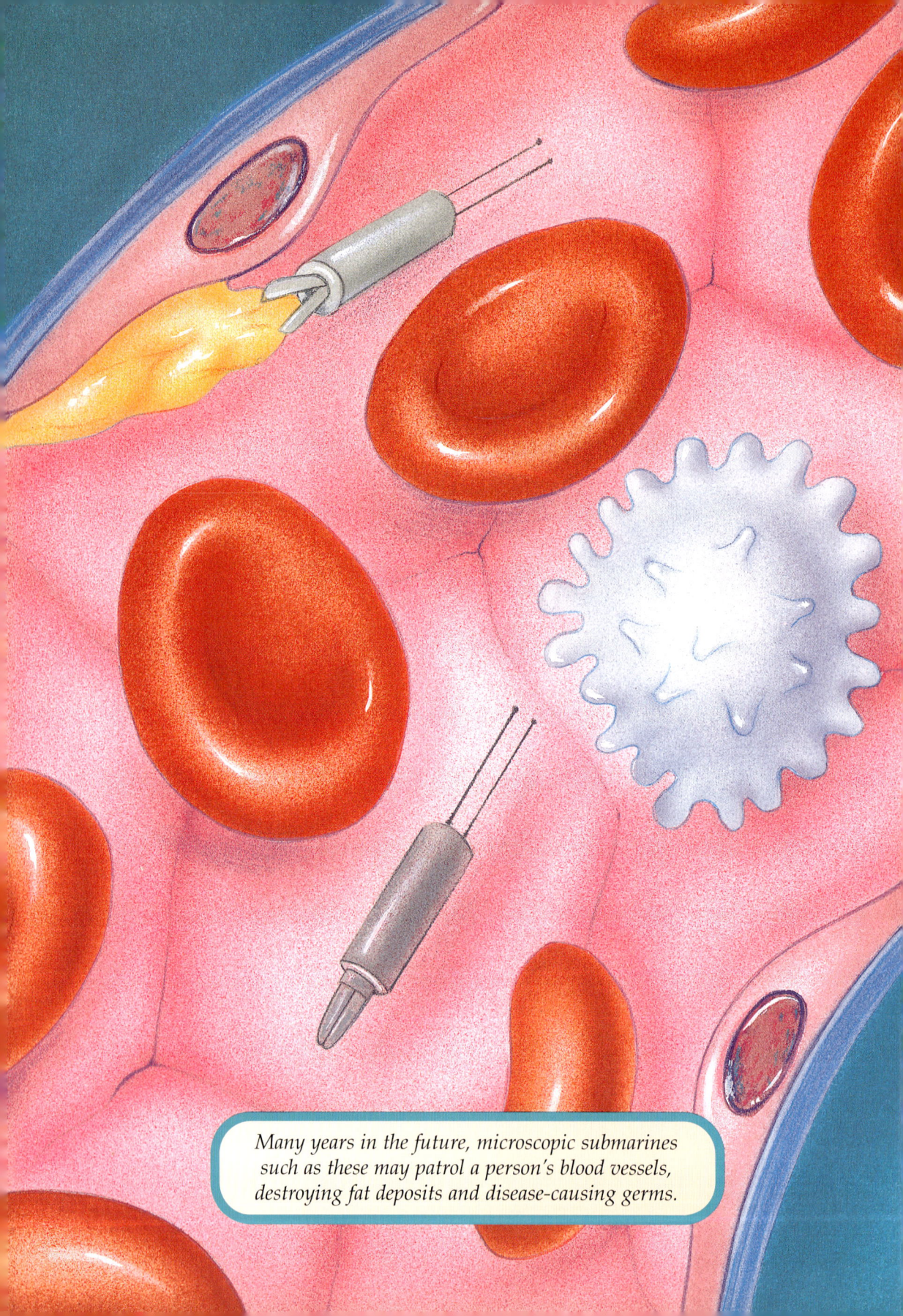

Many years in the future, microscopic submarines such as these may patrol a person's blood vessels, destroying fat deposits and disease-causing germs.

 # Introduction

Imagine that somewhere in your bloodstream is a tiny submarine. Its task is to patrol your arteries and veins and destroy any substances that might threaten your health. This amazing miniature sub is less than a millionth of an inch long and is powered by sugar and oxygen that it takes from your blood.

At the front of the vehicle are instruments for recognizing the kind of substance that lies immediately ahead. If these instruments sense "friendly" chemicals, such as those on the wall of the blood vessel or in the blood itself, the sub continues on its way. But if the object ahead is a hostile germ, the sub opens a set of mechanical jaws and swallows up the invader. Buildups of fat are attacked, too, before they can start to clog arteries. The unwanted substances are broken down inside the sub, made harmless, and released back into the bloodstream for removal by the kidneys. Now suppose there is not just one microscopic submarine but a fleet of billions and billions of them on constant patrol throughout your body.

Swarms of other tiny devices, far too small to be seen by the human eye, might help remove hazardous gases from the earth's atmosphere or digest garbage on waste dumps. Such remarkable machines would affect our lives in ways we can hardly begin to imagine. They may be among the eventual products of a new branch of technology called **nanotechnology.***

** Words that appear in **bold** are explained in the glossary on page 57.*

Introduction

The central, spinning part of this micromotor, which looks like a motorbike wheel, measures just 100 millionths of a meter (about 4 thousands of an inch) across.

At present, microscopic submarines are found only in Hollywood movies. However, researchers are making rapid progress in the design and construction of very small machines known as **micromachines**. In 1988 a team from the University of California at Berkeley built an electric motor that had turning parts finer than a human hair. Scientists are also learning how to isolate and control the positions of individual **atoms** and **molecules**—the tiny particles of which all matter is made. In 1990 American researchers created a message using 35 atoms of the rare gas xenon. They wrote their company's name, IBM, in letters that were 500,000 times smaller than the print on this page!

The study of micromachines and nanotechnology is so new that no one knows where it will eventually lead. All that seems certain is that it will have a big effect on people's lives in the twenty-first century.

In this book we will look briefly at the history of **miniaturization** and the reasons people have found for making machines smaller and smaller. We will look at the recent progress achieved in constructing tiny working devices and the limits nature places on the size of human-made things. Finally, we will speculate about some of the directions in which the technology of the ultrasmall may develop in the future.

CHAPTER 1

The Incredible Shrinking Machine

Why should we try to make machines, or the parts inside them, smaller and smaller? First, miniaturization allows them to be carried around or moved more easily. A machine that is portable can be used in many more situations than a big, heavy device. Second, if a machine is built from very small parts, it can perform highly complicated functions. Instead of ten working parts, it might have a thousand, or ten thousand, or more. Because of its complexity, it can do sophisticated tasks that would be impossible for a simpler machine. Some of today's machines, such as television sets, compact disk players, and industrial robots, work in ways that would have seemed almost magical to people living a few centuries ago.

If you were to take apart a device like a car, a personal computer, or a washing machine, you would find inside a huge collection of bits and pieces, precisely manufactured and fitted together. We take such complicated products for granted and rarely think about exactly how they work. But centuries of effort have gone into developing the techniques of miniaturization on which our world now depends.

Among the earliest masters of the very small were the Chinese. Elaborate toys with tiny working parts were built in China as long ago as 200 B.C. Over the next thousand years or so, all kinds of fantastic playthings, from mechanical flying birds to an otter that caught fish, were made for the amusement of the Chinese emperors.

Time in Your Hands

Intricate toys have often been used to show off a craftworker's skill. But in Europe, especially beginning about the sixteenth century, the desire to produce small, finely made components stemmed from more practical needs.

One of these needs was for people to be able to tell the time wherever they happened to be. Before 1500 all clocks were big and unwieldy because they used heavy weights to drive their machinery. Then a German locksmith invented the mainspring—a coiled ribbon of steel that, as it unwound, drove a series of cogs connected to the clock's hands. Being light and compact, the mainspring allowed timepieces to be built that could easily be carried around. This advance made possible the development of the pocketwatch and, eventually, the wristwatch.

The Incredible Shrinking Machine

Built by Pierre Jacquet-Droz, "The Scribe" uses two sets of interlocking wheels to control its writing hand.

In 1772 a Swiss clockmaker adapted the most advanced clockwork machinery of his day to make a lifelike doll that could write messages with a quill pen. The doll moved its hand almost with the precision of a real person.

Inventions Galore

By the end of the Middle Ages, people were finding all sorts of reasons for building intricate new machinery with small working parts. Navigators needed very accurate clocks, called chronometers, to tell the time, so they could work out their position at sea. Scientists needed precision-made microscopes, telescopes, balances, and other measuring equipment in order to carry out more detailed experiments.

As has often been the case in human history, war spurred the growth of new technologies. Gunmakers were urged to improve the accuracy and efficiency of firearms, for instance.

With the development of the steam engine in the eighteenth century, complex machinery could be driven to weave cotton and to propel steam trains. The invention of the telegraph, radio, television, and many other devices followed, along with the means to deliver electric power into people's homes.

When a new device is developed, it may not work particularly well. It is likely to be clumsy, hard to move and operate, and expensive to manufacture. Along with other improvements, engineers try to find ways to make the parts of a new invention smaller and lighter.

The Incredible Shrinking Machine

Small, finely-crafted parts allow this old pocket watch to keep accurate time.

This bulky computer of 1948 was many times less powerful than a modern personal computer.

Miniature Record Breakers

When it comes to making and measuring small things, there seems to be almost no limit to human ingenuity. For example, the world's smallest bicycle, belonging to a German circus performer, weighs only 1,050 grams (37 ounces) and has wheels that are just 4.5 centimeters (1.77 inches) high. Timekeeping technology has progressed even further. The smallest watch, made in Switzerland, weighs under 7 grams (0.25 ounces) and measures 1.2 centimeters (just over 1/2 inch) long and 0.476 centimeters (3/16 inch) wide.

The most advanced scientific balances in the world are capable of measuring to an accuracy of 0.00000001 of a gram—or less than 1/10 the weight of the ink on the dot over this *i*. The finest cuts have been made at the Lawrence Livermore National Laboratory in California. Using a special device, scientists were able to slice a human hair 3,000 times lengthwise.

It's a Small, Small World

To see how much progress has been made in miniaturization, compare the goods in a Sears catalog from around 1900 with the same kind of products available today. Think also about how much more a personal computer can do than one of the world's most powerful, room-filling "electronic brains" forty years ago.

We live in a world of advanced technology—made possible, in part, by our ability to manufacture very small, high-quality components. But just how small can human-made devices become?

CHAPTER 2

Computers, Crystals, and Chips

If you are sitting at home or at school reading this, you are probably surrounded by dozens, or perhaps even hundreds, of **microchips**. They are hidden away inside TVs, hi-fis, wristwatches, computers, calculators, kitchen appliances, and many other modern gadgets. They help to control the latest car engines and passenger aircraft.

A microchip is also known as an **integrated circuit**. It is made from a single crystal of a substance called **silicon**. The crystal is no bigger than your little fingernail and thin enough to pass through the eye of a needle. On the surface of this sliver of silicon, a complicated pattern of tiny electronic components and connecting pathways has been laid down.

Microchips are the most advanced products of miniaturization yet made. Their story began roughly a century ago.

A complete, special-purpose computer can be formed on a silicon chip smaller than an ordinary spider.

The Dawn of Electronics

When a piece of metal is heated in an empty space—a vacuum—the metal gives off a stream of particles called **electrons**. Think of electrons as being the tiniest possible packets of electricity. A flow of electrons, therefore, creates an electric current. If you imagine moving electricity to be like water flowing over a waterfall, electric **current** is equivalent to the amount of water that goes over the waterfall each second.

In 1904 an English scientist, John Fleming, found a simple way to control an electron flow. He invented the **diode**. Built from two pieces of metal, known as **electrodes**, inside a vacuum-filled glass tube, the diode let current pass through in only one direction. The flow of electrons from the first electrode, which was hot, could be stopped or started depending on the voltage on the second electrode. If you think again of a flow of electricity as being like a waterfall, then **voltage** is equivalent to the height of the drop.

In 1906 an American scientist, Lee de Forest, took Fleming's idea a step further with his invention of the **triode**. By placing a third, wire-mesh, electrode between the other two, de Forest was able to control the amount of current passing through the device. Very small changes in the voltage on the third electrode would cause big changes in current across the tube. This fact made triodes very useful as electronic components—devices that regulate a flow of electricity. Triodes were first used in radio transmitters. Later they found their way into radio receivers, phonographs (record players), and television sets.

Switched On

Scientists also realized that triodes could work as switches. In other words, they could be made to flip from "on" (when a current passes through them) to "off" (when no current passes through). This was very important because switches are the most vital parts needed inside computers. Sets of switches can be used to represent instructions and information.

In the late 1940s the first electronic computers were built using triodes. One of the earliest, called ENIAC, contained 18,000 triodes and weighed 30 tons! These computers were marvels of their age, but they suffered from some major problems.

Glass triodes and other large components, such as these from a computer of 1960, have now been replaced by small silicon chips, each containing millions of components.

Triodes are bulky, use a lot of power, and become hot when working. Computers made from them filled whole rooms, used a great deal of electricity, and had to be cooled by big fans. They were always breaking down, too, since each triode had a fairly short life.

Breakthrough

In 1947 three American scientists, William Shockley, John Bardeen, and Walter Brittain, invented an important new electronic device. It worked just like a triode but was solid, small, and needed very little power. The invention was a **transistor**.

To work as an electronic switch, a transistor did not have to be heated. Instead, it relied on the special electrical properties of a **semiconductor**. This is a substance whose ability to pass a current improves as tiny amounts of impurities, known as **dopants**, are added to it. The first transistors were built from a semiconductor called germanium. Eventually, however, silicon became the main semiconductor from which transistors and other electronic components were made.

Different parts of a crystal of silicon can be set up to act in the same way as the electrodes in a triode or a diode. This change is made by adding various tiny amounts of dopants to the silicon, such as boron in one region and phosphorus in another region. The result is a device that is small, stays cool, and lasts for a very long time.

In the 1950s transistors began replacing triodes as the basic building blocks of all computers. As a result, computers quickly shrank in size and became more dependable. Because they were smaller, information could travel around them more quickly, so computers also increased in speed.

At the same time, people began to realize all the different ways in which small, reliable computers might be applied. The space age had just begun, and computers were needed aboard spacecraft. But before they could be used for such tasks, computers would have to be reduced still further in size.

The Quarter-Inch-Square Marvel

Scientists began to ask: Instead of making components like transistors one at a time, why not make many of them together on the same crystal of silicon? The result would then be an integrated circuit—or a silicon chip.

In 1960 the first chip was made in the United States. Within a few years, chips containing several dozen transistors and other components were being manufactured. By 1975 the record stood at about thirty thousand components per chip. Because each component was now microscopic in size, the chips on which they were made became known as microchips. By 1994 it was possible to place over three million components on a single piece of silicon about a quarter-of-an-inch square!

In order to make a modern microchip, one draws a plan of all the components and connections. This plan is reduced in size hundreds of times, through a series of photographs, until it is only about five or ten times bigger than the actual chip. From the miniaturized plan of the chip, a template, or mask, is created.

Fire and Light

Wafers of pure silicon are sliced from a rod-shaped crystal and put into a red-hot furnace. The heat causes the wafers to produce a thin outer layer of **silicon dioxide**—a substance through which electricity cannot flow. Then the wafers are taken out of the furnace and coated with a type of chemical known as a **polymer**. The particular kind of polymer used in microchip manufacture is soft and sensitive to light.

A coated wafer is carefully positioned under the mask

containing the details of the chip to be manufactured. **Ultraviolet light** is shone through the mask and focused by an arrangement of lenses so that it forms a tiny, sharp image of the circuit pattern on the light-sensitive polymer. Ultraviolet light is used because the waves that make up ordinary light are too far apart to pick out the very fine details on a modern chip. In ultraviolet light the waves are closer together and so can produce a sharper outline through the mask.

After the first exposure, the wafer is moved to a slightly

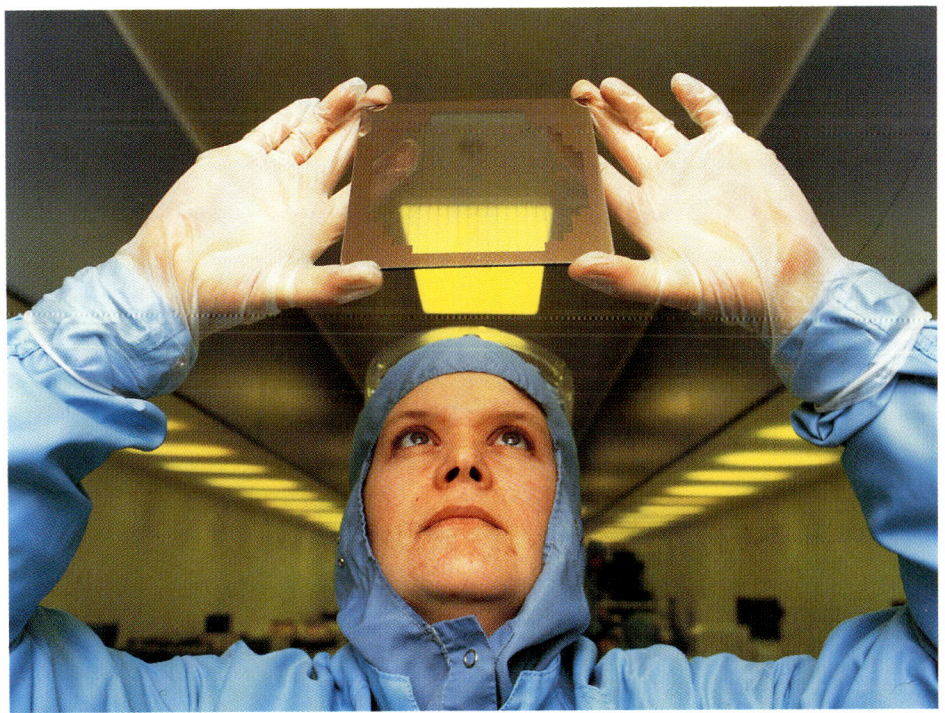

A worker checks a mask to be used in the manufacture of silicon chips.

different position so that another tiny region can be exposed to the ultraviolet light shining through the mask. In this way many copies of the chip are formed from one silicon wafer.

The Acid Test

The exposure of the polymer to light changes the polymer's solubility—the ease with which it can be dissolved by a special chemical. This chemical is known as a developer and is used to remove the polymer in the regions where it has been exposed.

The remaining polymer is baked to make it chemically resistant, and then the whole wafer is bathed in acid. Where the polymer has been dissolved, the acid is able to attack the underlying silicon dioxide and so expose the silicon below. As a result, the pattern of exposed silicon exactly matches the details contained on the mask.

Tiny amounts of different dopants are put onto the exposed silicon, at specific points. In this way, regions of the chip are formed that will work as diodes, transistors, and other components. Finally, very thin aluminum pathways are laid down to provide electrical connections.

Modern microchips may have ten or more different layers of components and connections. All the manufacturing stages just described must be repeated to produce each of the circuit layers. This means the chips must go back into the furnace to begin the whole process again and again.

Challenges of Tomorrow's Chips

For a number of years scientists have been asking: How much more can chips be miniaturized? Today's most advanced chips contain several million transistors. If progress continues at the present rate, by the end of this century, there will be chips containing about one billion transistors. At some point, researchers are likely to run up against problems that prevent a further reduction in size.

Already, microchips produce more heat for their area than the hotplate of an electric oven. Since the chips would fail if they become too hot, cooling systems are needed that can remove heat as fast as it is produced. In the future, a limit may be reached on how quickly excess heat from chips can be carried away.

Another problem arises from excessive electrical effects. The voltage across a single transistor on a chip is quite small, but it occurs across such a tiny distance that it amounts to an extremely steep voltage drop. Such a voltage drop produces a powerful electrical effect in its immediate neighborhood. As the components on a chip are crowded closer and closer together, the electrical effects they give rise to may become so large that they prevent the transistors from working properly.

Many other difficulties confront the chip manufacturers of tomorrow. Perhaps the biggest challenge of all will be to find ways to create the ever-tinier components and pathways on the chip's surface. Its fineness of detail is limited by the size of

Computers, Crystals, and Chips

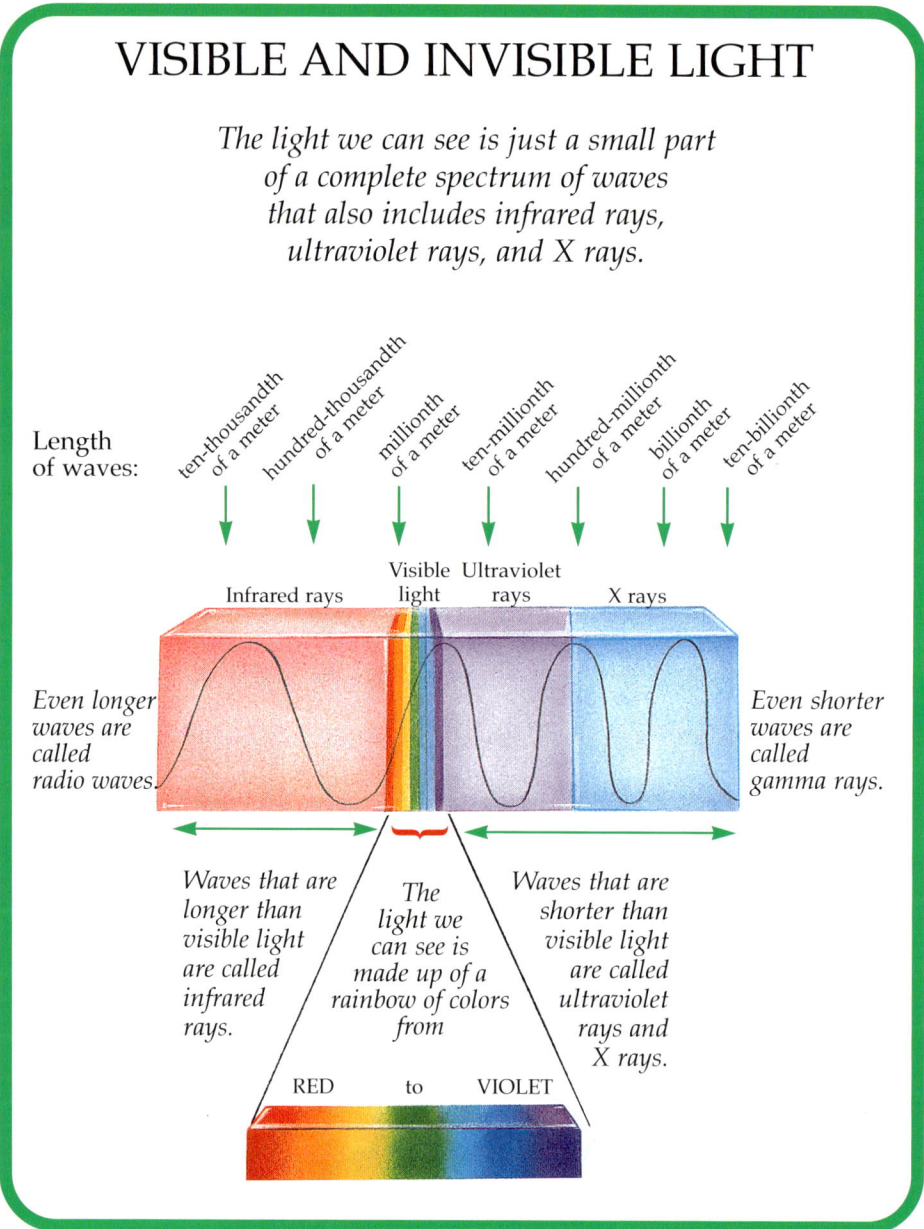

the waves that are used to shine on the mask. Ordinary light has already been replaced by ultraviolet light. Now even the limits of ultraviolet rays are being reached, and some scientists believe that the next generation of integrated circuits will have to make use of **X rays**, which consist of even shorter waves than ultraviolet.

Limits to the miniaturization of chips may eventually be reached, but not, scientists believe, before you will be able to carry a computerized copy of all the books in the Library of Congress around in your pocket.

SUPERSPEED SILICON ?

How can computers and other electronic devices be made even faster and more powerful than they are today? One way is by miniaturizing the components on a silicon chip still further. This would reduce the distances that signals have to travel and so increase the speed with which information can be handled. An alternative approach is to look for new materials through which electrical signals can move faster.

A basic limit to the speed of a computer is the rate at which its transistors can flip from "off" to "on" and back again. Some scientists are investigating materials in which this so-called switching speed is much higher than it is in pure silicon. One of the new possibilities they are looking at is an alloy of silicon and germanium. New transistors made from a silicon-germanium alloy have been shown to switch on and off about three times faster than the fastest silicon transistor ever made. Researchers are now experimenting with ways to incorporate these superspeed switches into practical circuits.

CHAPTER 3

Gearing Down

Researchers have found out how to make electronic components incredibly small. Now they are also learning how to create microscopic versions of machines like gears, pumps, and motors. These miniature moving gadgets are known as micromachines or micromechanical devices.

Many people have heard about microchips and how important they are in all sorts of ways. By comparison, not so much has been said in the media about micromachines. One of the reasons for this lack of attention is that micromachines are a lot harder to make and so scientists are still working to perfect them. Over the next few years, however, micromachines will become more and more common in everyday life. By early in the next century, they may play just as important a part in the running of our modern world as microchips do today.

Microscopic Motors

Micromachines took a giant step forward in 1988 when a team from the University of California at Berkeley demonstrated a working electric motor that could be seen only with the help of a microscope. This motor had a main spinning part, or rotor, that measured just 60 **micrometers**, or millionths of a meter, across. If this rotor had been made ten times bigger, it would still have been only as wide as a pin.

Since then, even smaller micromotors have been built. Teams of scientists all over the world are studying how to make these motors work more efficiently and produce more turning force, or **torque**, so that they will be able to do useful jobs.

One of the biggest problems with micromotors is **friction**, the force that acts when surfaces rub against one another. Friction is what stops your bicycle when you put on the brakes. We depend on friction to prevent us from skidding every time we walk. But inside a machine that is supposed to move freely and easily, friction is a nuisance.

With an ordinary-size electric motor, such as the one in a washing machine, the frictional forces usually add up to less than a hundredth of the torque that the motor produces. In micromotors, however, the friction can be as big as the torque. When this happens, the motor is stopped from moving.

The first micromotors tended to lock up after just a few thousand spins. Scientists working on these very small devices realized that they would have to pay special attention to

A tiny mite, barely visible to the unaided eye, moves toward a micromechanical device.

designing bearings that cut down friction as much as possible. Their efforts have been very successful. The latest micromotors are still hampered by friction, but the amount of friction can be reduced to less than 10 percent of the torque. As a result, micromotors can now run for days on end at low speeds or in short bursts at up to 15,000 revolutions per minute.

Making a Micromachine

Using methods borrowed from silicon-chip technology, researchers started making micromachines in the 1970s. But they quickly realized that these methods would have to be adapted in various new ways, because microelectronic components are more or less flat, whereas micromechanical ones are three-dimensional. Micromachines have length, width, *and height*.

To construct a micromachine, scientists make a multilayerer silicon sandwich. On the bottom is a single crystal of silicon. Above this is a layer of silicon dioxide. Next comes another layer of silicon, called polycrystalline silicon, or **polysilicon**, because it is made of lots of little jumbled-up crystals. On top of this is a second layer of silicon dioxide, followed by a second layer of polysilicon.

The moving parts for a micromachine are made by masking and dissolving away specific regions of the second (top) polysilicon layer. Acid is used to remove the unwanted portions of the silicon dioxide layer below, leaving the moving parts free to spin or pivot. The underlying polysilicon layer provides

Gearing Down

HOW TO MAKE A PART FOR A MICROMACHINE

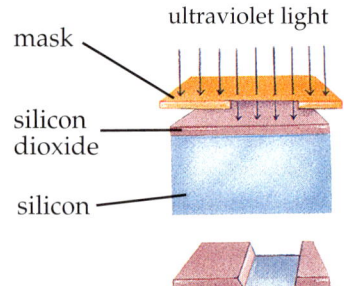

(1) Ultraviolet light is shone through a mask onto a layer of silicon dioxide.

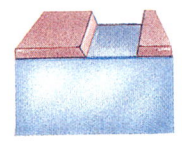

(2) Those parts of the silicon dioxide which were exposed to the ultraviolet light are removed with acid.

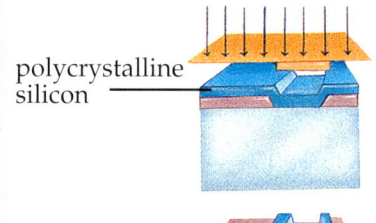

(3) A layer of polycrystalline silicon is laid down and ultraviolet light shone through a second mask.

(4) Where the ultraviolet light shone through, the polycrystalline silicon is removed.

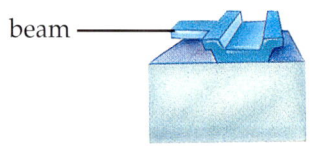

(5) The rest of the silicon dioxide is removed with acid, leaving a part for a micromachine—in this case, an overhanging beam.

the secure base upon which the micromachine rotors and levers can move.

Alternatives to Silicon

In some ways silicon is a good building material for micromachines. It is almost twice as hard as iron and more difficult to stretch than steel. Silicon's biggest drawback is that it is brittle—it can crack or shatter quite easily. Even little bits of silicon that break off inside a micromachine can act as grit that stops the device from working.

In the early 1980s a group of scientists at the Karlsruhe Nuclear Research Center in West Germany began developing a new method for making micromachines. This method is known by the initials—**LIGA**. Using LIGA, researchers can build micromachines from **nickel** and other metals that are much less brittle than silicon.

Instead of the ultraviolet rays presently used in microchip technology, LIGA uses X rays. The X rays are shone through a mask—a stencil of the machine being made—onto a layer of polymer. Then a developer is used to dissolve the polymer where the X rays have penetrated it. Up to this point, LIGA follows steps similar to those involved in making silicon microchips.

Next, however, an electrical process fills the gaps in the polymer with a nickel or a metal like nickel. When this has been done, the rest of the polymer coating is dissolved away.

Gearing Down

The nickel shape that is left behind can be used either directly as part of a micromachine or as a master mold from which to make copies. To produce copies, technicians place a layer called a casting plate over the nickel. Through tiny holes in this casting plate, hot liquid plastic is injected. When the plastic has cooled and hardened, it is lifted off and used as a mold for making duplicates of the original nickel structure.

Better Micromachines

To make a micromotor with spinning parts that are finer than a human hair is an amazing achievement. But can these tiny devices do any useful jobs? One of the problems with micromotors is that they produce very little torque. Once the friction between their moving parts has been overcome, they can often barely keep their own rotors spinning around, much less help turn any larger devices to which they might be attached.

One way to increase the power of a micromotor is to build it less like a pancake, because the size of a motor's torque increases with the height of its spinning parts. The methods used for manufacturing silicon micromachines are not suited to producing structures that are higher than about 10 micrometers. However, if nickel and the LIGA process are used, this height limit can be overcome.

Researchers in Germany have made a nickel device that is only 5 micrometers wide, but 300 micrometers high. At the University of Wisconsin, another team of scientists has made toothed nickel gears, 50 to 200 micrometers across and 200 to

Micromachines and Nanotechnology

300 micrometers high. These toothed wheels are linked together to form gear trains. Such gear trains could eventually be used to transmit the motion of a micromotor to other pieces of equipment nearby.

MICRO VELCRO

Researchers at Carnegie Mellon University in Pittsburgh have shown that microchip-manufacturing methods can be used to create what they call micromechanical Velcro. First they grow a silicon dioxide layer on a silicon base. Then they repeatedly mask chosen areas and remove the unwanted parts with acid. The process produces silicon dioxide arrowheads, or hooks, on silicon pedestals. Each structure is no more than 18 micrometers wide, and 200,000 of them can be packed onto a square centimeter.

The tiny arrowheads will pierce human flesh and then hold fast because of their barbed ends. The Pittsburgh team believes the arrowheads could be used to rejoin severed blood vessels and to close wounds in place of stitches. But unlike everyday Velcro, the microscopic version does not pull apart easily. Instead, the hooks tend to break off as they are separated

Scientists at Carnegie Mellon University in Pittsburgh used microchip manufacturing methods to create what they call "micromechanical Velcro." Two hundred thousand of these tiny barbs can be packed onto a square centimeter.

Gearing Down

Swimming Bacteria and Medicine

In Japan a team of scientists is investigating another kind of micromechanical device that could eventually be used in medicine. These scientists are looking at ways to propel machines inside a patient's body to deliver lifesaving drugs. Ordinary micromotors might not be able to move in blood, which is quite a thick liquid. So, to solve this problem, the Japanese scientists are looking at the possibility of copying the tiny whiplike structures that some kinds of bacteria use to move around. These structures, which are only 30-billionths of a meter in diameter, are called flagella.

It has been known for some time that parts of a flagellum work like an ordinary electric motor. Close to where it joins the main body of the bacterium, the flagellum has a rotor that spins around inside a fixed ring.

The Japanese scientists want to discover the exact arrangement of chemicals that help such a tiny structure to work so effectively. With this knowledge, they hope to make a computer model of the flagellum and then try to figure out how to produce an artificial version. Such an advanced micromachine probably lies many years in the future, however.

CHAPTER 4

How Small Can You Get?

Take a grain of sugar and break it in half. Now take one of these smaller grains and break it in half again. Suppose you have the means to keep on splitting the bits of sugar exactly in two, again and again, even after they are much too tiny to be seen. Eventually you will reach a limit. You will end up with a fragment of sugar so small that if you divide it just one more time it will no longer be sugar. This smallest possible piece of sugar is called a molecule.

Most natural and human-made substances are made of molecules. Molecules, in turn, are built of even smaller particles called atoms. One molecule of household sugar (sucrose) consists of 12 atoms of carbon, 22 atoms of hydrogen, and 11 atoms of oxygen, joined together. A simpler molecule, such as that of water, contains just 2 atoms of hydrogen and 1 of oxygen.

How Small Can You Get?

ATOMS AND MOLECULES

All matter is made of atoms. There are 92 different kinds of atoms in nature, from hydrogen, the simplest, to uranium, the most complex.

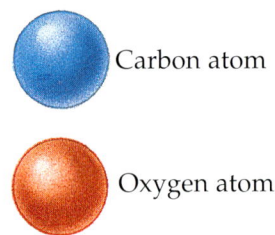

*Atoms join together to make molecules.
For example, two atoms of oxygen and one atom of carbon make one molecule of carbon dioxide.*

In diamond, which is a form of carbon, each carbon atom joins with four others in a tetrahedron (a figure with 4 triangular faces).

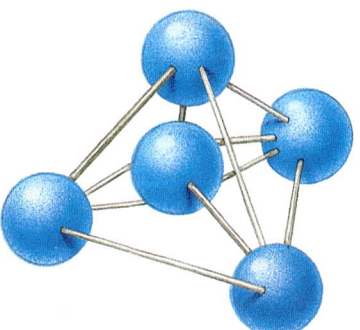

Some substances, such as the precious metal gold and the lightweight gas helium, consist of atoms that are not linked together to form molecules. So, with the right equipment, you could pluck a single atom of gold out of a gold ring or a single atom of helium out of a helium-filled balloon.

Sizing Up Atoms

It is very hard to visualize how small atoms really are. Ten million of them placed in a line would fit across the period at the end of this sentence. Put another way, if everything in the world were blown up in size so that a single atom were as big as this period, then the smallest letters on this page would be about 50 miles high!

Even the smallest of ordinary things is made up of unimaginably huge numbers of atoms. A single pin, for instance, contains roughly a million times as many atoms as there are human beings on the earth.

An ordinary hand-held machine, such as a hair dryer, contains about as many atoms as there are grains of sand on all the world's beaches. When a machine like this is built, no one knows or cares where each individual atom inside is located. It might seem beyond belief that people could eventually learn how to build anything one atom or molecule at a time.

Nature's Building Blocks

Atoms are incredibly small, but we know without a doubt

How Small Can You Get?

The flat faces of this crystal of quartz are caused by the atoms inside being lined up in a regular way.

that it is possible to put them together, one by one, in a very orderly way. We know this because nature does it all the time. The perfectly flat sides and regular shapes of crystals are due to the precise arrangement of atoms inside the crystals. As a crystal forms, individual atoms add on to the growing structure like people sitting down, one at a time, from front to back, on the neat rows of evenly spaced seats in a theater.

The same orderly growth of structures takes place within living things. A human being is not a confused heap of atoms and molecules. As a baby grows, it does so according to a definite pattern. Like all the different kinds of animals and plants, a human being assembles itself in an organized way from individual atoms and molecules.

If our bodies can control their growth at the atomic and molecular level, then perhaps scientists and engineers can learn to do the trick by artificial means. This suggestion was first made by the brilliant American scientist Richard Feynman, in a speech in 1959. Feynman said: "Consider the final question as to whether, ultimately—in the great future—we can arrange atoms the way we want; the very *atoms*, all the way down! What would happen if we could arrange atoms one by one, the way we want them?"

Nanotechnology Is Born

If we could find a way to put atoms together, one at a time, exactly as wanted, we could build almost anything at all. We could make a robot spacecraft that was smaller than a pollen grain. We could mass-produce a billion of these tiny explorers and scatter them throughout space, just as real pollen is blown about on the wind. In the remote future, marvelous things like these may be possible. But first, researchers will have to become experts at working with atoms.

Some years after Richard Feynman made his speech about

building with atoms, the author Eric Drexler coined the name *nanotechnology*. People have been familiar for a long time with words beginning with **micro**. There are *microscopes, microbes* (tiny living things), and *microchips*, to name a few. *Micro* comes from the ancient Greek *micros*, which means "small." In science, *micro* also stands for 1-millionth. A micrometer, for instance, is 1-millionth of a meter.

For defining even smaller measurements, scientists use the prefix *nano*. This comes from the Greek *nanos*, meaning "dwarf." A **nanometer** is a thousand times smaller than a micrometer, or 1-billionth of a meter.

Because *micro* was a term already used to refer to anything microscopic, *nano* is a good name to describe the possibility of working at even smaller scales. Nanotechnology, then, is the technique of building machines and other structures from individual atoms and molecules.

INSIDE THE ATOM

Given enough energy, atoms themselves can be split apart into even smaller particles. At the center of an atom is the nucleus—a cluster of particles made up of protons and neutrons. The proton has a single positive charge, while the neutron is electrically neutral. Around the nucleus is a cloud of even smaller particles called electrons, each of which has a negative charge that is equal in size and opposite in sign to the charge on the proton. Because, in every atom, the number of protons is the same as the number of electrons, atoms have no overall charge.

Electrons are believed to be pointlike specks of matter that cannot be broken apart any further. Protons and neutrons, however, are each thought to be a kind of bag containing three smaller particles known as quarks.

Scientists are now learning how to control the movements of individual electrons or small groups of electrons. This knowledge may eventually allow them to create new kinds of computer memories—for example, the storage of fantastic quantities of information. For the purposes of nanotechnology, however, the smallest possible building blocks will be atoms and molecules.

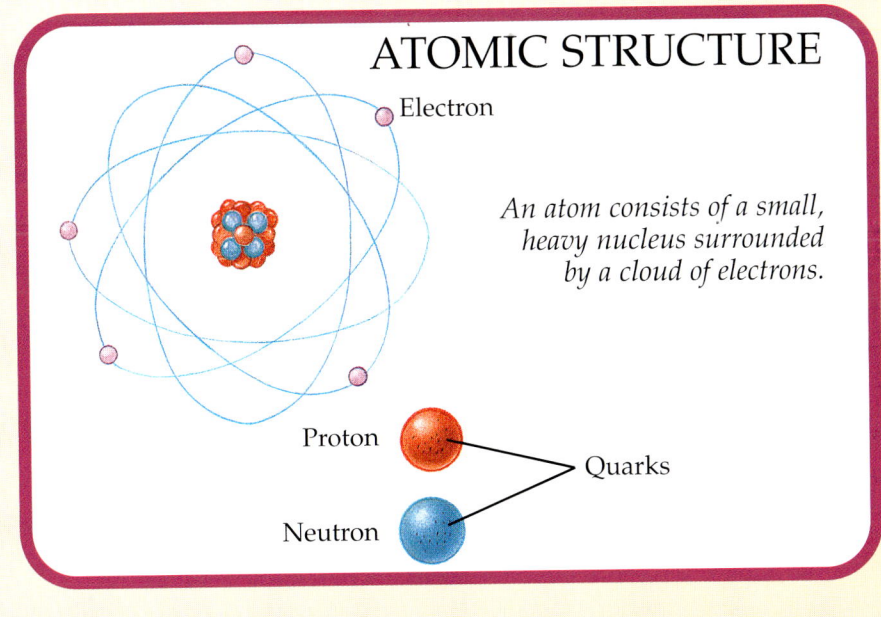

ATOMIC STRUCTURE

An atom consists of a small, heavy nucleus surrounded by a cloud of electrons.

CHAPTER 5

The Atomic Erector Set

Nature manages to put structures like crystals together one atom or molecule at a time. But how can human beings work with building blocks that are this small? Until recently, the very idea would have seemed incredible. But within the past few years, scientists have developed means to pick up individual atoms and move them around with great precision. By applying this new skill, they can draw pictures or write words by placing atoms exactly where they want them.

Seeing the Very Small

Before you can build with individual atoms, you have to be able to see where they are. Ordinary microscopes, like those you may have used at school or at home, are known as optical microscopes because they work with visible light. They are ideal for studying objects the size of either small pond creatures or of crystals of salt and sugar. But a microscope that uses visible light cannot possibly show anything as minute as an atom or molecule.

A microscope is limited by the **wavelength** of the radiation that it employs. The wavelength is just the distance between the crest of one wave and the crest of the next. Visible light has a wavelength of between 400-thousandths and 700-thousandths of a centimeter. Short though that may seem, it means that the smallest object an optical microscope can reveal is still many times larger than an atom.

A scientist uses an electron microscope to magnify objects up to 40,000 times.

The Atomic Erector Set

In the 1930s, scientists developed a much more powerful instrument called the electron microscope. As with all of the smallest particles in nature, electrons can behave both like tiny bullets of matter and like waves. Electron waves are much shorter than light waves so that electron microscopes can detect more detail in whatever is being examined. However, even an electron microscope cannot give a clear view of objects the size of atoms.

Supermicroscopes

To see individual atoms, scientists needed a completely new type of microscope. The first of these superpowerful instruments was built in 1981; it is called the scanning-tunneling microscope.

All solid objects are covered with an incredibly fine cloud of electrons. This cloud is thickest and rises farther from the object's surface wherever there is an atom, just as real clouds tend to hang around mountaintops. Using a scanning-tunneling microscope, a researcher can map the electron cloud on a tiny patch of a surface and so find out exactly where the atoms below are located.

The most important part of a scanning-tunneling microscope is its needle, or probe, made from the metal tungsten. This needle is made very sharp by various chemical and electrical processes. At its tip, it is so sharp that there is just a single atom or very small cluster of atoms sticking out above the rest.

Micromachines and Nanotechnology

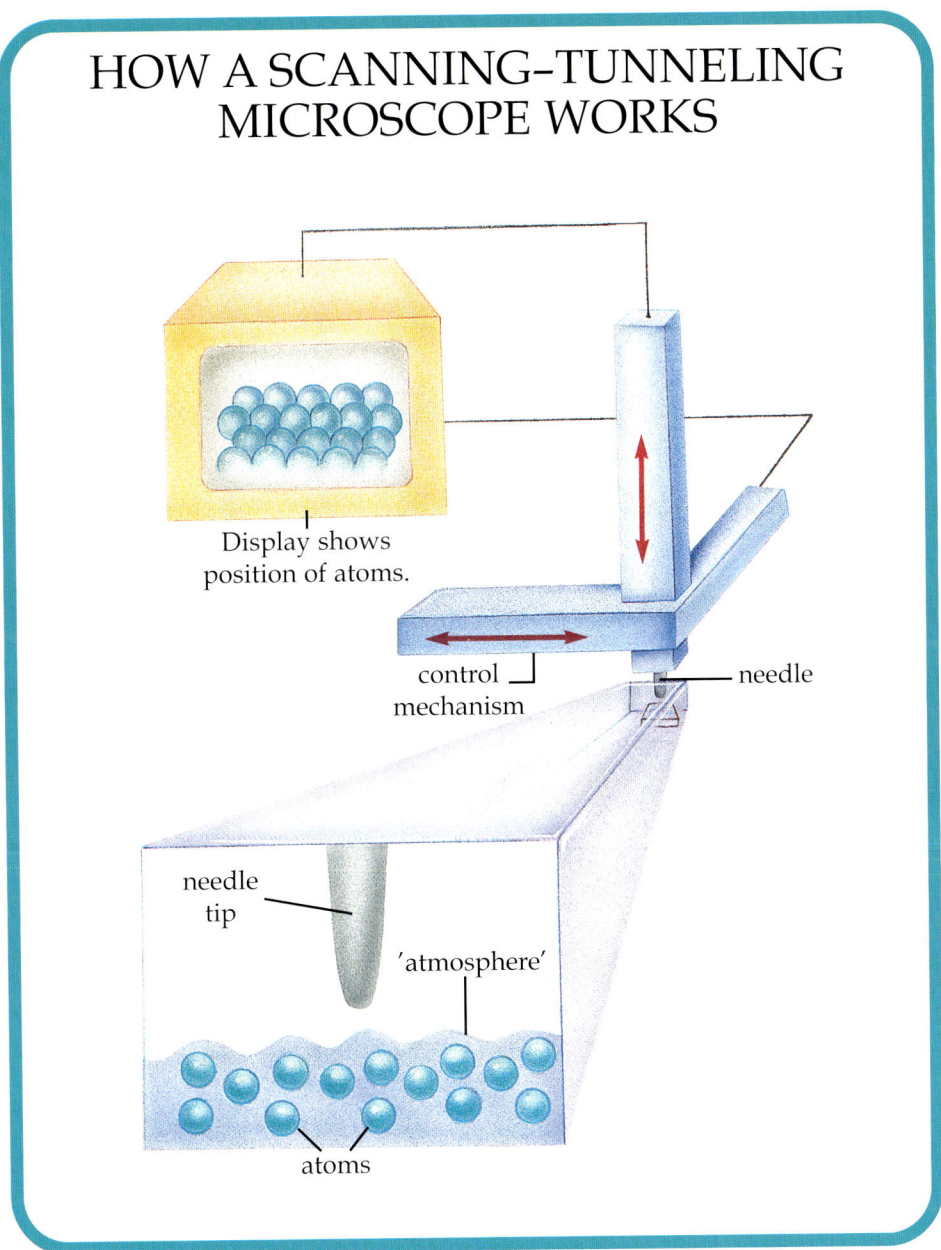

The Atomic Erector Set

This ultrafine needle is moved slowly, back and forth, across the surface being examined, advancing by less than a nanometer after each sweep. Although the needle never actually touches the surface, it comes so close that the electron cloud covering the surface and the electron cloud surrounding the tip of the needle overlap. A minute voltage is applied to the needle. This allows electrons to hop from the surface to the needle, a movement that results in a tiny current of electricity. By just the right amounts to keep this current the same, a sensitive control system raises and lowers the needle as it moves along.

The up-and-down movements of the needle are turned into a picture on a television screen that shows where each individual atom or molecule on the surface is located. Roughly speaking, a "mountain" on the picture corresponds to an atom, a "valley," to where there is no atom.

Putting Atoms in Their Place

Scientists realized that the scanning-tunneling microscope could do more than just show atoms. It could also be used to pick them up and move them around.

Normally the voltage on the tip of the microscope's needle is kept very low, to avoid disturbing the atoms on the surface. But if the voltage is increased when the needle is placed exactly over an atom, then that atom can be made to stick to the needle tip. In 1990 two scientists at the computer company IBM used this technique to write their company's name one atom at a time.

Micromachines and Nanotechnology

The IBM researchers began by spraying atoms of the gas xenon onto a clean nickel surface that had been cooled to minus 269 degrees Celsius (minus 452 degrees Fahrenheit). At this very low temperature, the atoms remained very still. Using a scanning-tunneling microscope, the researchers found out exactly where the xenon atoms had landed on the surface. Then they put the needle over one of the xenon atoms, increased the voltage on the needle so that the atom stuck to the tip, and slid the atom across the nickel surface to the exact position they wanted it. When the atom was in place, the researchers dropped the voltage so that the atom was released. They repeated this process until 35 xenon atoms spelled out the letters *IBM*.

A year later the same team created *CO Man*. This was a picture of a little man made from 28 molecules of the gas carbon monoxide (which has the chemical formula CO).

Writing words and drawing pictures with atoms and molecules may seem like a waste of time. But the ability to place individual atoms and molecules precisely may prove to be very important in the future. Using this technique, scientists might build, atom by atom, custom-designed molecules with special properties. Other possibilities include the creation of atomic-scale data storage, so that, for instance, you might someday be able to hold all the information in all the world's libraries in the palm of your hand. In fact, comical creations like CO Man really mark the birth of nanotechnology.

The Atomic Erector Set

> **READING THE SMALL PRINT**
>
> It is surprising how tiny some people have been able to write even without the aid of advanced scientific instruments like the scanning-tunneling microscope. In 1926 a man used a diamond point connected to a mechanical device called a pantograph to put the Lord's Prayer onto an area of glass measuring just 0.04 millimeter (0.0016 inch) by 0.02 millimeter (0.0008 inch). In 1983 Tsutomu Ishii of Tokyo managed to write the names of 184 countries on a single grain of rice and the words *Tokyo Japan* in Japanese on a human hair!

MEET MOLECULE MAN

A team of researchers at IBM made this picture from 28 molecules of carbon monoxide. The molecules were put into place using a scanning-tunneling microscope. From head to toe, "molecule man" measures just 5 millionths of a millimeter!

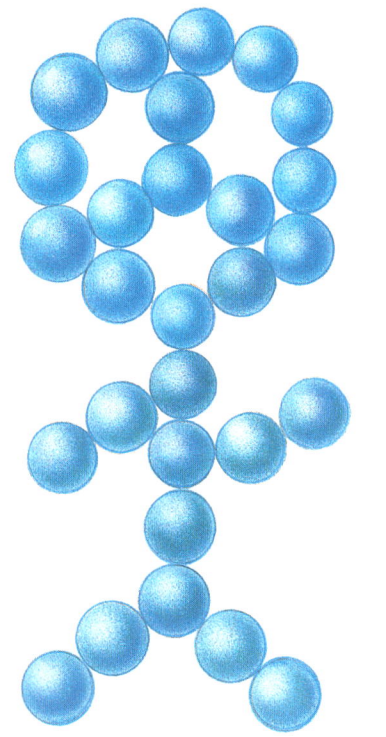

CHAPTER 6

Amazing Nanotechnology

In 1959, when Richard Feynman suggested the possibility of building structures one atom or molecule at a time, the idea seemed fantastic. Even Feynman underestimated how rapid future progress in miniaturization would be.

Feynman offered a $1,000 prize to the first person who could make a working electric motor that would fit inside a cube that measured 1/64 of an inch on each side. He guessed it would be a long time before he had to pay up. But just two and a half months later, William McLellan, a physicist at the University of California Institute of Science and Technology, claimed the prize. Working on his lunch breaks with a microscope, toothpick, and watchmaker's lathe, McLellan assembled a motor that met Feynman's requirements.

As we have seen, in the years since then, scientists have

William McLellan's prize-winning electric motor, which would fit inside a cube one sixty-fourth of an inch across, is seen next to a gnat's wing.

made enormous progress in developing new techniques of miniaturization. They have also taken great strides in other fields, such as learning how the chemicals inside our bodies work to help us grow and stay alive and healthy. In the years ahead, the dream of nanotechnology may come true as researchers put together the knowledge and methods from areas such as chip technology, micromechanics, and biochemistry (the study of the chemicals of life).

Nanomachines of the Future

Nature has been manufacturing complicated products from individual molecules for almost 4 billion years—for as long as there has been life on Earth. Your body is made of trillions of microscopic, living units called cells. Each cell is like a miniature factory that builds complex chemicals from relatively simple raw materials. The plans for making these chemicals are stored inside a special molecule called DNA. Other molecules act as messengers, carrying the instructions from DNA to sites within the cell, where the essential chemicals needed for life, including proteins, are put together.

What nature can do, it may be possible for human beings to copy. Perhaps in the future scientists will create "assemblers." These would be small machines that would take in raw materials from their surroundings and piece them together, atom by atom and molecule by molecule. They would be programmed to carry out certain construction tasks, turning out specific nanomachines by means of miniature assembly lines.

Every hour, entire factories no larger than a grain of sand might generate billions of machines that would look like a mass of dust streaming steadily out from the tiny factory doors. The purpose of these nanomachines would be determined by the programmed instructions of the factories that made them. They might be devices for cleaning up oil spills or other forms of pollution. They might be miniature submarines that could be injected into a person's bloodstream to destroy cancer cells or attack fat deposits.

The possible uses of nanomachines stretch our imagination to the limit. They could, for instance, be used to manufacture food in any desired amounts, wherever and whenever it was needed. Scientists have even suggested the possibility of a nano "meat machine" that would create fresh meat from raw materials in the environment. There could also be nano "egg machines" and "wool machines," so removing the need to keep animals at all for human use.

> ### A Slippery Problem
>
> How do you keep a nanomachine running smoothly? A car engine has to be well lubricated with oil; otherwise the friction between its moving parts quickly causes the engine to overheat and seize up. But the methods of lubricating an ordinary-size machine would not work with a device that was only as big as a large molecule. At this scale, the effects of individual molecules of oil sliding against one another would become overwhelming—the oil would seem as sticky and sluggish as tar.
>
> Fortunately, it might not be necessary to lubricate bearings in nanomachinery at all. As Richard Feynman pointed out, such tiny bearings could be allowed to run dry since any heat they produced would escape almost immediately.

A World Transformed

We are still many years away from seeing these astonishing developments. However, it would be a mistake to suppose that they could not happen at all. What to one generation might seem unbelievable—or unacceptable—can seem commonplace to the next. Microchips and scanning-tunneling microscopes would have appeared incredible 50 years ago. But now they are a reality.

The development of powerful machines of ever smaller size promises to bring great changes to the world in which we live.

The prospects for nanotechnology are extraordinary. Just as today we are surrounded by countless numbers of tiny creatures that live on us and inside us, so our world may eventually include trillions upon trillions of artificial molecule-sized machines.

These nanomachines could be used for much that is good. They could keep our environment clean, mopping up pollutants as soon as they are released. They could do hazardous jobs, such as working underground or in nuclear reactors, that today have to be carried out by people. They could help us to live longer and healthier lives.

But, as with almost all forms of technology, there is the possibility for misuse. Just as an army of micromachines could be programmed for constructive purposes, so it could be turned to terrible ends. Perhaps more seriously, there is the possibility that nanomachines might be given the wrong instructions by mistake. For instance, an army of such devices intended to clean up pollution might cause further damage to the environment instead. This raises a serious question. If malfunctioning nanomachines did go on the rampage, how would we ever find them so that we could stop what they were doing?

We stand on the brink of a world that may be transformed by devices that we cannot even see. One way or another, nanotechnology promises to be among the most far-reaching developments of the twenty-first century.

 # Glossary

atom—The smallest particle of an *element*—a substance made up of only one type of atom. Examples of elements include iron, oxygen, carbon, and silicon.

current—The quantity of electricity flowing around a circuit.

diode—A very simple electronic component. Depending on the voltage of its two electrodes, a diode can pass or block a flow of electricity.

dopant—A substance used to alter the electrical properties of a semiconductor.

electrode—A piece of material that conducts electricity and allows an electric current to enter or leave a device.

electron—A particle found inside atoms. Electrons carry a single, tiny electrical charge. A lot of electrons moving together gives rise to a flow of electricity.

friction—The force that acts when any two surfaces come into contact and slide against one another. Friction can occur with solids, liquids, and gases.

integrated circuit—An arrangement of electronic components and connections manufactured on the surface of a single crystal of silicon. It is also known as a silicon chip or microchip.

Glossary

LIGA—A fairly recent technique developed for making micromachines. LIGA uses X rays to reproduce the very fine details of a micromachine and an electrical process to create the machine's parts from a metal such as nickel.

micro—A prefix, or beginning of a word, that comes from the ancient Greek term for "small." A microscope, for example, is used to look at very small objects. In scientific units of measurement, "micro-" stands for 1-millionth.

microchip—A silicon chip that contains many microscopic components.

micromachine—A microscopic-sized machine. The parts of such a machine may measure between a few micrometers and a few hundred micrometers across.

micrometer—One-millionth of a meter; also known as a micron. About 25,000 micrometers equal one inch.

miniaturization—The process of making things smaller in size.

molecule—A group of atoms joined together in a particular way. In most substances, a molecule is the smallest portion that can take part in a chemical reaction.

nano—A prefix, or beginning of a word, that comes from the ancient Greek term for "dwarf." In scientific units of measurement, "nano-" stands for one-billionth.

nanometer—One-billionth of a meter.

nanotechnology—The technique of building structures and machines from individual atoms and molecules.

nickel—A silvery white metal that is similar to iron. Nickel is a popular material for building micromachines because it can be laid down in very thin layers by a method called electroplating.

polysilicon—A form of silicon consisting of large numbers of small crystals.

polymer—A substance, either natural or artificial, consisting of long-chain molecules.

semiconductor—A substance whose ability to conduct electricity increases with rising temperature, and with the presence, in small quantities, of substances known as dopants.

silicon—A nonmetallic element that occurs as dark gray crystals. It is the most widely used semiconductor for making integrated circuits.

silicon dioxide—A substance that occurs naturally as sand. It can be formed as a layer by heating silicon to a high temperature. Unlike silicon, it will not conduct electricity at all.

torque—The total twisting force about a pivot.

transistor—A very important electronic component. The amount of electricity passing between two of the electrodes of a transistor can be accurately controlled by the voltage on a

Glossary

third electrode. This property allows transistors to be used either as control components or as switches.

triode—A device, made from metal electrodes inside a vacuum-filled glass tube, which is used to control a flow of electricity. Today, triodes have been largely replaced by transistors.

ultraviolet light—Light that is invisible to the human eye. It consists of waves that are shorter than those of violet light.

voltage—The force with which a flow of electricity is pushed around a circuit.

wavelength—A measure of the distance between the tops of two successive waves in a series of waves.

X rays—Similar to the light that we can see but made up of much shorter waves. X rays consist of waves that are even shorter than those of ultraviolet light.

For Further Reading

Asimov, Isaac. *How Did We Find Out About Atoms?* (How Did We Find Out About . . . series.) Walker & Co., 1976. An easy introduction to the building blocks of matter and how scientists have gradually learned more about them.

Asimov, Isaac. *How Did We Find Out About Robots?* Walker & Co., 1984. From the same series, a history of the development of robots. This includes a description of the earliest automatons and the tiny, intricate parts needed to make them work.

Bender, Lionel. *Invention.* (Eyewitness Books.) Knopf, 1991. A beautifully illustrated guide to some of the most important inventions in history including many (such as clocks, microscopes, and integrated circuits) that pushed back the frontiers of miniaturization.

Cicciarella, Charles F. *Microelectronics and the Sports Sciences.* Human Kinetics, 1986. An explanation of how electronic measuring equipment is built for research in the sport sciences.

Evans, Christopher. *The Micro Millennium.* Washington Square Press, 1982. An entertaining look into the future role of microcomputers.

Articles:

For more advanced students the following articles on micromechanics and nanotechnology may be of interest.

"Frothing a Raindrop," *Scientific American*, May 1992, page 84.

"Micron Machinations," *Scientific American*, November 1992, pages 106–117.

"Scanned-Probe Microscopes," *Scientific American*, October 1989, pages 74–81.

 # Index

assemblers 53
atoms 8, 37, 38, 39, 41, 42, 43, 44, 45, 46, 48, 49, 51, 57, 58

Bardeen, John 20
Brittain, Walter 20

Carnegie Mellon University 35
cells 52, 53
China 11
CO Man 49
computers 15, 16, 18, 19, 20, 21, 27

de Forest, Lee 19
diode 18, 21, 24, 57
Drexler, Eric 42

electrodes 18, 19, 57, 59, 60
electron cloud 46, 48
electrons 18, 43, 46, 48, 57
ENIAC 19

Feynman, Richard 41, 51, 54
flagella 36
Fleming, John 18
friction 29, 31, 34, 57

IBM 9, 48, 49

integrated circuits 17, 22, 57, 59
Ishii, Tsutomu 50

Jacquet-Droz, Pierre 12

Karlsruhe Nuclear Research Center 33

Lawrence Livermore National Laboratory 15
LIGA 33, 34, 58

mainspring 11
McLellan, William 51, 52
medicine 36
microchips 17, 22, 24, 25, 28, 33, 54, 57, 58
micromachines 8, 9, 28, 29, 30, 31, 32, 33, 34, 58
micromotors 8, 29, 31, 34, 35, 36
microscope, electron 45, 46
microscope, optical 44, 45
microscope, scanning-tunneling 46, 47, 48, 49, 50, 54
molecules 9, 37, 38, 39, 41, 42, 43, 48, 49, 51, 52, 53, 58

nanomachines 52, 53, 54, 55, 56
nanotechnology 7, 9, 42, 49, 52, 55, 56, 59

semiconductor 21
Shockley, William 20
silicon chips 20, 22, 23, 27, 57

Index

torque 29, 31, 34, 59
transistor 20, 21, 22, 24, 25, 27, 59
triode 19, 20, 21, 60

University of California at Berkeley 8, 29
University of California Institute of Science and Technology 51
University of Wisconsin 34

Velcro 35

watches 11, 14, 15